Measuring

David Kirkby

RIGBY
INTERACTIVE
LIBRARY

© 1996 Rigby Education
Published by Rigby Interactive Library,
an imprint of Rigby Education,
division of Reed Elsevier, Inc.
500 Coventry Lane,
Crystal Lake, IL 60014

Cover designed by Herman Adler Design Group.
Designed by The Pinpoint Design Company
Printed in China

00 99 98 97 96
10 9 8 7 6 5 4 3 2 1

Library of Congress Cataloging-in-Publication Data
Kirkby, David, 1943–
        Measuring / David Kirkby.
        p.   cm. — (Mini math)
        Summary: A simple introduction to the
mathematical concepts of length, weight, quantity, and
time, with some easy activities included.
        ISBN 1-57572-004-3 (lib. bdg.)
        1. Mensuration—Juvenile literature.
[1. Measurement.   2. Time.]   I. Title.   II. Series:
Kirkby, David, 1943– Mini math.
QA465.K52   1996
530.8—dc20   3644   95-38706
                                                CIP
                                                AC

Acknowledgments
The publishers would like to thank the following for the
kind loan of equipment and materials used in this book:
Boswells, Oxford; The Early Learning Centre; Lewis',
Oxford; W.H. Smith; N.E.S. Arnold. Special thanks to the
children of St Francis C.E. First School.

Photography: Chris Honeywell, Oxford

Contents • Conten

tallest

shortest

This doll is tallest.

This doll is shortest.

4

Which is tallest?

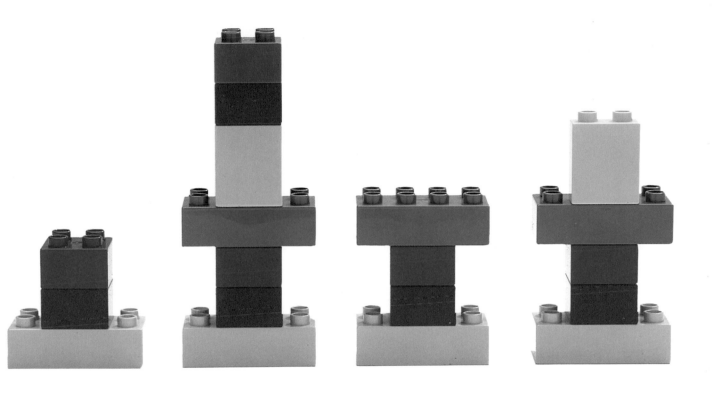

1   2   3   4

• To Do •

Copy the towers above.
Now draw some more
bricks on tower 1
to make it the tallest.

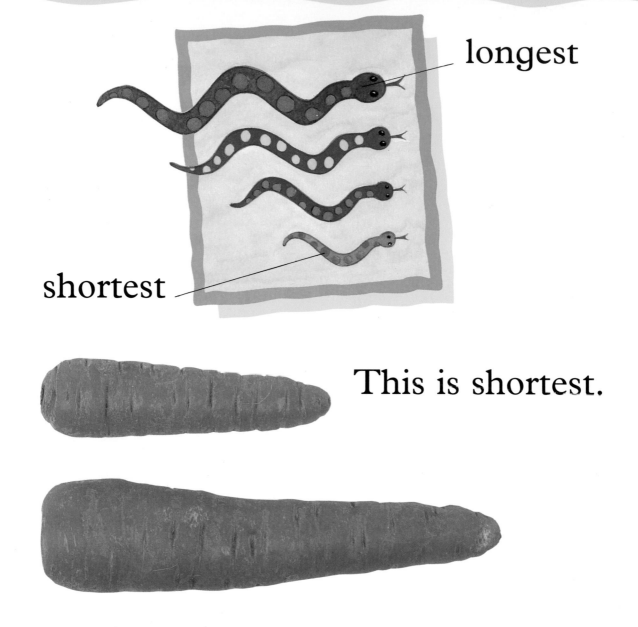

longest

shortest

This is shortest.

This is longest.

6

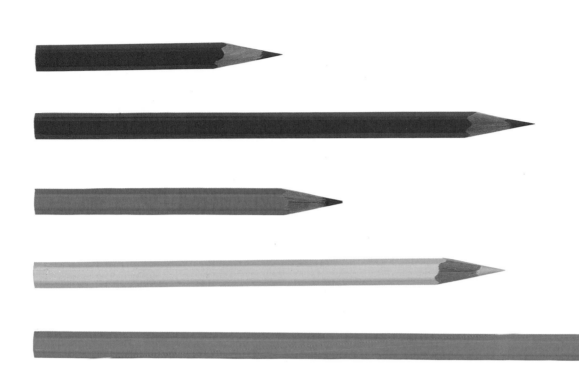

Which is longest?
Which is shortest?

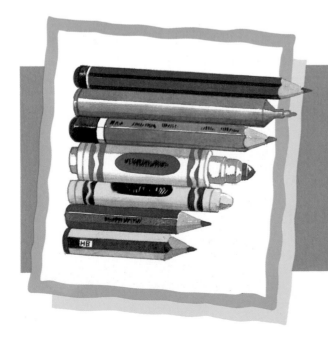

• To Do •

Find some pens and pencils. Line them up. Put the longest at the top. Put the shortest at the bottom.

7

Sometimes you need to know how long or how tall something is.

You can measure with anything.

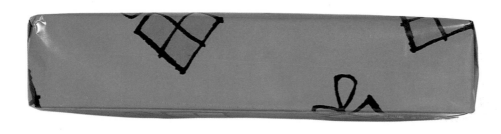

Measure with the same thing.
This present is 9 pasta tubes long.

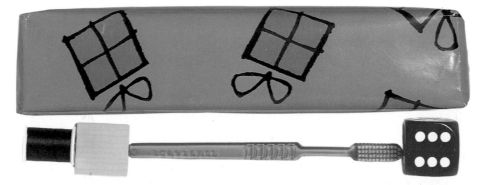

It is hard to measure well with different things.

8

How many boxes tall are the girls?

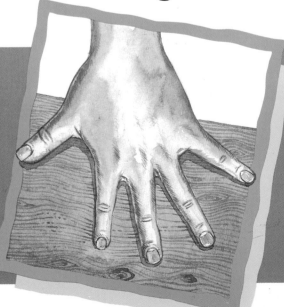

• To Do •

Use your open hand to measure with.
How many hands long is your table?
How many hands tall is your table?

It is best if everyone uses the same thing to measure.
We measure in inches and feet.
12 inches = 1 foot.

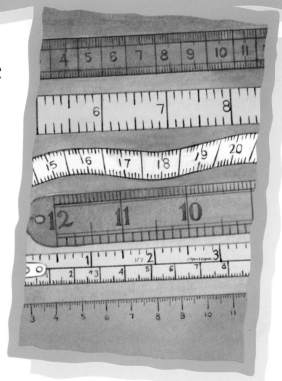

Kieran is 53 inches tall.

He is 24 inches around.

How many
inches tall?
How many inches wide?

• To Do •

Estimate (guess) how
many inches long
your foot is.
Measure it.
Estimate (guess) how
many inches it is
around your head.
Measure it.

11

We use scales to
weigh things.
They tell us
which is heavier.
The heavier thing
makes the scales go down.

The dinosaur is heavier.
The ball is lighter.

Which is lighter?

• To Do •

Name 5 things that are lighter than a bed.
Name 5 things that are heavier than a ball.

13

We weigh things in ounces and pounds.
16 ounces
= 1 pound.

The candy weighs 1 pound.

1 pound

1 pound

The potatoes weigh 1 pound.

How much does this flour weigh?

• To Do •

Find out which weighs more: a box of crackers or a bag of flour.

The pitcher holds more juice than the glass. We say that the capacity of the jug is greater than the capacity of the glass.

6 glasses of juice will fill this pitcher. The capacity is the amount of juice the pitcher will hold.

16

We measure liquid capacity in cups, pints, and quarts.

• To Do •

Estimate (guess) which will hold more water: a glass or a bowl.

Fill a glass and a bowl with water to see if you were right.

Clocks measure time in minutes and hours. The long hand is the minute hand. The short hand  is the hour hand. When the long hand is on 12, it is the beginning of a new hour.

This clock shows 8 o'clock.

One clock shows a different time
from the others.
What time does it show?
What time do the other clocks show?

The weather is

• To Do •

What time is it?

Digital clocks do not have hands. They show the time just using numbers. This clock shows 8 o'clock. The number before : shows the hour. The number after : shows the minutes. When the clock shows **:00**, it is the beginning of a new hour.

This clock shows 10 o'clock.

What time is it?

• To Do •

Write these times down as digital time.

We also measure time in days, weeks, months, and years.

7 days = 1 week. 52 weeks = 1 year.
12 months = 1 year.

Monica is 6 years old today.

How old will she be
in 12 months' time?

• To Do •

When is your birthday?
How many months old
are you?

Page 5	Tower 2 is tallest.
Page 7	The green pencil is longest.
	The red pencil is shortest.
Page 9	5 boxes tall
Page 11	8 inches tall, 6 inches wide
Page 13	The bag is lighter.
Page 15	The flour weighs 8 ounces.
Page 19	1 clock shows 6 o'clock.
	The others show 2 o'clock.
	To Do: 4 o'clock, 9 o'clock, 12 o'clock
Page 21	10 o'clock
	To Do: 8:00, 10:00, 5:00
Page 23	7 years old